Ada
LoVELacE

Ada
LoVelacE

by Nancy F. Castaldo

Illustrated by Charlotte Ager

Editorial Assistant Katie Lawrence
Designer Charlotte Jennings

Senior Editors Marie Greenwood, Roohi Sehgal
Project Art Editor Roohi Rais
Jacket Coordinator Issy Walsh
Jacket Designer Dheeraj Arora
DTP Designers Sachin Gupta, Mrinmoy Majumdar
Project Picture Researcher Sakshi Saluja
Pre-Producer Sophie Chatellier
Senior Producer Ena Matagic
Managing Editors Laura Gilbert, Monica Saigal
Deputy Managing Art Editor Ivy Sengupta
Managing Art Editor Diane Peyton Jones
Delhi Team Head Malavika Talukder
Creative Director Helen Senior
Publishing Director Sarah Larter

Subject Consultant Dr. Christopher Hollings
Literacy Consultant Stephanie Laird

First American Edition, 2019
Published in the United States by DK Publishing
1450 Broadway, Suite 801, New York, NY 10018

A catalog record for this book is available from the Library of Congress.
ISBN: 978-1-4654-8540-3 (Paperback)
ISBN: 978-1-4654-8541-0 (Hardcover)

DK books are available at special discounts when purchased in bulk for sales promotions,
premiums, fund-raising, or educational use. For details, contact:
DK Publishing Special Markets, 1450 Broadway, Suite 801, New York, NY 10018
SpecialSales@dk.com

Printed and bound in China

A WORLD OF IDEAS:
SEE ALL THERE IS TO KNOW

www.dk.com

Dear Reader,

Ada Lovelace grew up in the early 19th century in a world that had few female mathematicians or scientists. In addition, her mother worked hard to limit her curiosity and keep her head out of the clouds. She was keen to prevent Ada from becoming too much like her unpredictable father, the Romantic poet, Lord Byron. These factors may have discouraged Ada, but they didn't—her curiosity was too strong to be crushed. Despite all attempts, her imagination was so vivid that she was able to use it to see a future with technology unlike that of her own time. She never stopped asking "What if?"

It is because of her unrelenting curiosity and vision in imagining what computers might accomplish, even before they existed, that I am able to type this book on my computer and chat on my smartphone. Imagine if Ada was around to see her thoughts spring to life. Never stop dreaming and asking "What if?" Like Ada, we all have the power to make a difference.

Dream on, readers!

Nancy F. Castaldo

The life of...
Ada
Lovelace

1
WATCHING RAINBOWS IN THE SKY
page 8

6
MEETING CHARLES BABBAGE
page 50

7
LIKE WEAVING FLOWERS
page 58

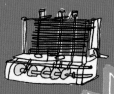

8
THE COUNTESS OF LOVELACE
page 66

9
THE ANALYTICAL ENGINE
page 76

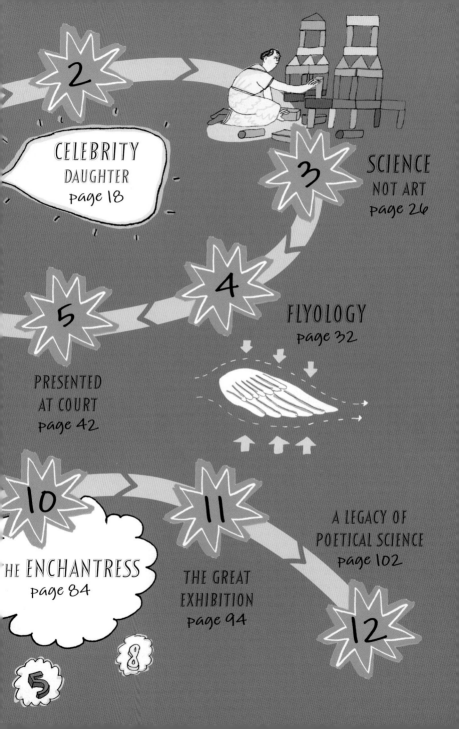

2

CELEBRITY DAUGHTER
page 18

3
SCIENCE NOT ART
page 26

5

4
FLYOLOGY
page 32

PRESENTED AT COURT
page 42

10

11
A LEGACY OF POETICAL SCIENCE
page 102

THE ENCHANTRESS
page 84

THE GREAT EXHIBITION
page 94

12

5

8

WATCHING **rainbows** IN THE SKY

As a girl, Ada was clever, curious, and inventive. Later, these traits helped her to be thought of as the world's first computer scientist.

Ada was the daughter of celebrities—famous from the day she was born. Her father was the legendary poet, Lord George Gordon Byron, and her mother was Lady Anne Isabella "Annabella" Milbanke, a woman who adored math.

Annabella was worried that her daughter would grow up to be foolish and unpredictable

like her famous father—with good reason. Lord Byron's world was filled with chaos. He liked to gamble and he had many love affairs. Unable to live with Byron any longer, Annabella took baby Ada and went to live with her parents.

Annabella did not want Ada's imagination to run free, and she wanted to make sure that Ada shared her love of math. Annabella told the people who looked after Ada to only speak the truth to her. She tried so hard to keep Ada from thinking about fantastical, nonsensical things, but, being curious, Ada wondered about them anyway.

Byron also thought that Ada should focus on the certainty of science. He wrote to Annabella when Ada was growing up to find out more about her personality and what interested her.

Byron inquired whether she was talkative or quiet, shy or sociable, and passionate and imaginative. He also really hoped that Ada

Ada Byron

would not become a poet like him. Annabella sent a miniature portrait of their daughter to Byron. This came with a letter that told Byron that Ada was a cheerful girl. Annabella also said that Ada used her imagination, but only while thinking about mechanical objects, such as ships. Creativity would prove to come in handy in Ada's later life, when she needed to solve complex math problems.

Byron died before being able to send a reply. Ada didn't really know her famous father, but his large character and reputation were always present in her life.

---

what does reputation mean?

The opinion that people have about a person, or to be known for something. Byron had a reputation as a great poet.

"**I hope** that the Gods have made her **anything** save poetical – it is **enough** to have **one such fool** in a family."

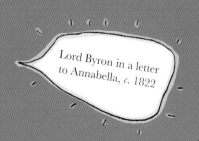

Lord Byron in a letter to Annabella, *c*. 1822

Byron might have been unpredictable, but he had other qualities that made him an exceptional poet. Like her father, Ada was inventive and always observing. These traits would help her to achieve great things when she grew up. As a young girl, Ada watched birds to figure out how they were able to fly. She wondered about the clouds in the sky, and was intensely interested about everything she saw and everywhere she went.

No matter how hard Annabella tried, she couldn't squash Ada's curiosity. Ada showed a strong desire to understand how things work. She was especially curious about rainbows, as she wanted to discover the science behind them, not just admire their beauty. Ada spent a lot of time studying rainbows, and noticed that if you look at the sky after it rains you might see one. Sometimes, Ada looked closer and saw a second rainbow. To find out why this happened, she wrote to her tutor, William Frend.

William Frend

Ada wanted to know why all the rainbows she had seen were curve-shaped, why they seemed to form part of a circle, and how second rainbows are made. She instinctively knew how the colors of a rainbow are separated, but could not grasp why the colors appear differently when there are two rainbows in the sky.

William had also tutored Ada's mother, Annabella. He was a traditional academic who taught his students "certainty, not uncertainty," only wanting to focus on scientific fact.

HOW ARE RAINBOWS MADE?

When the sun is behind you, and it's raining in front of you, you might be able to see a rainbow. Rainbows can be seen when the sun's white light shines through raindrops. This white light is split into an arc of different colors. If you look closely at a rainbow, you'll see seven colors: red, orange, yellow, green, blue, indigo, and violet.

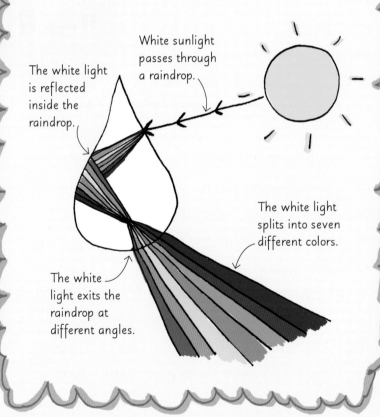

White sunlight passes through a raindrop.

The white light is reflected inside the raindrop.

The white light splits into seven different colors.

The white light exits the raindrop at different angles.

William only wanted to teach Ada about things that were certain, so he agreed to answer her questions about rainbows.

It's likely that he would have explained that the second rainbow is caused by a double reflection of sunlight inside raindrops—rather than the single reflection of sunlight inside raindrops that makes a single rainbow.

The angle of light from the double reflection means that the second rainbow looks like it's upside down. The colors go from violet on the outside to red on the inside, the opposite of a normal rainbow!

William's explanation should have satisfied Ada's curiosity about rainbows, but she always had more questions about other topics. Whether she asked her tutors these questions, or later found her answers in books, Ada never stopped wanting to know more about how things worked.

Ada lived during a time when scientific study was not encouraged in girls. However, she did not let that, or her mother's attempts to discourage her, stop her search for answers. Ada's thoughts had no bounds—she was always determined to learn as much as she could.

We know that Ada's curiosity and endless questions led her to imagine one of the most important inventions of our world—the programmable computer.

19TH-CENTURY WOMEN OF SCIENCE

In the 19th century, it was not easy being a woman who wanted to learn about science. For one thing, there were few places for them to study. The well-known universities in England—Oxford and Cambridge—only let men attend class. Women could only learn from books and private tutors. Three women who did not let their gender stand in the way of their scientific curiosity were Mary Anning, Florence Nightingale, and Maria Mitchell. Mary became a fossil collector, who discovered important dinosaur bones. Florence chose to become a nurse instead of marrying a wealthy man. Her success in the field of nursing improved cleanliness in hospitals and saved countless lives. Maria became an astronomer and discovered a comet.

Mary Anning

Florence Nightingale

Maria Mitchell

2

Celebrity **daughter**

To understand Ada's complicated childhood as the heiress daughter of her celebrity parents, let's look more closely at them.

Ada's mother, Annabella, was an only child who lived in the English countryside with her wealthy parents. She grew up at the end of the 18th century—a time when many people thought marrying a wealthy man was the most important thing for a young woman to do. Annabella's family also believed that it was very important for her to have an education, not for a career, but

Annabella

What is an heiress? A woman, such as Ada, who will inherit wealth, property, and status from her family or other person.

because it would serve her well in marriage.

Many tutors arrived from Cambridge University to teach her philosophy, science, literature, languages, and math. Annabella was especially talented at math, a subject studied mainly by men at the time.

Ada's father, Byron, also grew up in England in a wealthy family. His father, Captain John Byron, died when Byron was around four years old. Although he was just a little boy, he became an English lord at 10 years old when his great uncle, William Byron, passed away. He inherited the title Lord Byron and gained William's properties, high status in society, and money.

Lord Byron

Byron was sent to boarding school, then to college, and he began to spend a lot of money. In fact, he spent more money than he had—running into a lot of debt.

Byron's wild behavior spread into every part of his life. There was even a rumor that he kept a tame bear in his college rooms at Cambridge University! Despite this chaotic lifestyle, Byron wrote poetry—and it was good. People took notice of him, especially after he made a powerful speech in the House of Lords in 1812 defending the Luddites in their protest against technology. Soon after his speech, some of his poetry was published and he was skyrocketed to fame.

Like most celebrities, everyone wanted him at their parties. Byron was a rockstar poet, who had lots of fans!

THE LUDDITES

The Luddites were a group of English workers in the 19th century. They protested against new technology coming into the textile mills where they worked, as this meant they were losing their jobs. Byron chose to defend the Luddites in his first speech in the House of Lords (one part of the British government). Byron's fame partly came from his protest against technology, yet Ada's fame would be linked to innovations in technology.

Annabella was the complete opposite of Byron. She lived a quiet life, studied hard, and was religious. When she was older she was sent to London to find a husband from high society—but she did not find one right away. She rejected potential husbands during her first two social seasons. The third season arrived and she attended Lady Caroline Lamb's dance party. This is where she first laid eyes on Byron.

What is a social season?

The time of the year, usually during winter, where members of high society, such as Annabella, can meet people they might marry.

Byron attended party after party in London, and Annabella was warned about his bad reputation. She fell head over heels in love with him anyway. After all, Annabella thought that Byron was handsome and interesting, and he was the bachelor of the year.

Byron also fell head over heels for Annabella. He fondly nicknamed the math whiz the "Princess of Parallelograms." The two were married in January 1815. Byron was 26 years old, and Annabella was four years younger.

Annabella brought a lot of money with her into the marriage—perfect for Byron with his debts and crazy spending habits. However, Annabella found Byron to be too unpredictable, and she thought he made a lot of bad choices.

WHAT IS A PARALLELOGRAM?

A parallelogram is a type of four-sided shape. Its opposite sides are parallel to each other. This means that they are the same distance apart along their whole length. A rectangle is a type of parallelogram. Byron's nickname for Annabella came from this shape, as she was so good at math.

This side is parallel to the bottom side.

This side is parallel to the left side.

The two were very different people and their marriage was doomed from the start.

That December, a daughter was born—Augusta Ada Byron, named after Byron's half sister, Augusta Leigh. By this time, Annabella knew that Byron was never going to change. She decided that his behavior was not good for her or baby Ada. It was time for them to separate. Annabella and Ada left Byron and their home in London.

Augusta Leigh

As soon as the celebrity couple split, gossip flew around the English high society. Byron fled to Italy, and although Ada's parents kept in touch through letters, Byron didn't see Annabella and Ada again.

Despite this, Ada still grew up in the spotlight. Annabella was determined that Ada would not grow up to be anything like her father. But of course, Ada inherited some of his traits, both the good and the bad.

Even though Ada did not know her father, he continued to be a huge influence on her life.

Ada as a child. She
is holding a bracelet
that has Byron's
portrait on it.

Chapter 2

Science not art

Ada lived with her mother and grandparents at Kirkby Hall in Leicestershire, England. There were rarely other children for her to play with.

When Ada was growing up, she was looked after by nannies and tutors. They made sure Ada followed strict rules that were set by her mother and grandmother.

Without many playmates, days were often lonely for Ada. This changed when she was five years old—Annabella gave her a Persian kitten. Ada adored the fluffy kitten and named her Mistress Puff. Annabella had a habit of firing tutors and nannies if they became close to her daughter and stopped being strict with her. But

Mistress Puff was always there to keep Ada company.

Soon after getting Puff, a new nanny arrived from Ireland to tutor Ada—Miss Lamont. She wrote that Ada was "brim full of life, spirit, and animation." Her new pupil was eager to learn about everything.

Miss Lamont's joy in teaching Ada was not matched by Annabella. Ada's mother demanded that her daughter's head and heart be planted firmly on the ground, similar to the lime trees that lined the carriage trail to their house.

Remembering her disastrous year of marriage to the wild Lord Byron, Annabella set strict rules for her daughter's lessons. Ada was taught lots of subjects, such as music, French, and math.

Ada was rewarded with tickets if she sat still in lessons—if she collected 12, they could be traded for a book Ada liked. If she did not follow Annabella's orders, she received punishments, such as being locked in a closet. Ada was punished a lot, but her lively spirit never went away.

Ada's parents wrote to each other about their daughter. Byron had a few requests for Ada's education. Although neither of her parents had any musical talent, Byron wanted Ada to learn music. He also wanted his daughter to be taught languages. When Ada was eight, she wrote a letter to her mother about a boy she had met called Hugo. He only spoke Italian and Spanish, but she could understand him.

THE ART OF LETTER WRITING

In the early 19th century, there were no phones, emails, or texts. Instead, people wrote letters. There was an art to letter writing—there was even a manual that explained the different types of notes, letters, and cards.

"The **little boy** is a very nice child on the whole, he speaks nothing but **Italian** and Spanish, which I now **perfectly understand**."

Ada in a letter to Annabella, c. 1824

Miss Lamont continued to teach Ada and followed Annabella's strict rules as best as she could. She really enjoyed tutoring Ada, as Ada was very clever—but the young girl didn't like all her subjects, least of all math. Annabella on the other hand loved math and made her daughter sit through her lessons. Ada struggled to concentrate, and her thoughts ranged far and wide.

When Annabella was away from Kirkby Hall, Ada's tutors let her explore. She was allowed to use her colored bricks to build cities and imagine distant lands during geography lessons. She wondered about everything, even what the waves in the sea looked like in faraway countries, such as Norway.

When Annabella returned to Leicestershire, Ada had to follow her strict rules once more. To keep Ada from moving her fingers during

her lessons, Annabella asked a maid to
wrap Ada's fingers with black cotton bags.
Ada bit the maid and was sent to her room
as punishment. Annabella was very angry
with her daughter.

After a while, Ada was allowed to go back
into the drawing room, and Annabella finally
calmed down. Ada's fingers
were unwrapped and
Annabella read poetry
to her daughter.

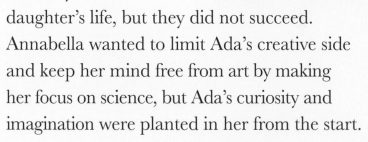

Annabella may
have introduced these
rules to try and bring
certainty to her
daughter's life, but they did not succeed.
Annabella wanted to limit Ada's creative side
and keep her mind free from art by making
her focus on science, but Ada's curiosity and
imagination were planted in her from the start.

Flyology

Like many wealthy young girls in England in the 19th century, Ada was taken on a whirlwind tour of the Continent.

After Lord Byron passed away, Annabella felt free to travel out of England with Ada. She decided to take the 10-year-old Ada on a special trip, crossing the English Channel and traveling around Europe. They would visit important cities and see famous sites. In the end, Ada and Annabella toured the Continent for 15 months.

It was a wonderful time for Ada. She wrote about the steamboats she saw, drew chalk sketches of the mountains in Switzerland, and

CONTINENTAL EUROPE

During the 19th century, "the Continent" referred to all of the countries in mainland Europe, excluding any islands. That meant that some places, such as Britain, Ireland, Sicily, Sardinia, and Cyprus, were not considered to be part of the Continent. Grand tours of the Continent could last many months.

enjoyed music and concerts. Ada and Annabella traveled in style. After all, Ada was the daughter of a very famous celebrity. During the trip they stayed in fancy hotels and ate expensive food.

As Ada grew, so did her imagination. A few months after they returned to England, Ada decided she wanted to fly just like a bird. She had observed a lot of birds, watching their feathered wings beat in the air above her. Ada thought it would be wonderful to fly just like them—one problem: she didn't have wings! Ada was determined to make her own wings, so she began to explore all the different materials that she could use, such as feathers, silk, and paper.

Ada wrote a letter to her mother dated April 2, 1828, to tell her that she had figured out how to make wings that could be attached to her shoulders. She told Annabella that she knew exactly how the wings would work and move. She also said that she wanted to make the

HOW DO BIRDS FLY?

Birds are a natural engineering marvel. Their lightweight, smooth feathers reduce air pressure above their wings, which creates an upward force called lift. Thin, hollow bones keep the wings lightweight and make flying easier. Birds have a streamlined, slim body, and strong muscles, while their powerful wings pull them into the air.

The difference in air pressure below and above the wing creates lift.

Air pressure is lower above the wing.

Air pressure is higher under the wing.

wings out of silk, and if that didn't work, she would try using feathers instead.

In the same letter she asked her mother for a book about birds, so she could better understand how they fly so well.

Letter after letter, Ada wrote to her mother about birds, their anatomy, and her fantasy of flying one day.

Ada also imagined a machine that could help her fly without wings: a steam-driven, mechanical flying horse. She thought about this project every day. It was hard to think of anything else—she desperately wanted to fly.

Ada outlined her plans for this flying horse to Annabella. She told her mother that she would first perfect her birdlike wings and then move onto a more elaborate invention.

What is anatomy? The study of the body of a human or animal.

Ada's flying horse would be powered by a steam engine. She declared that it would be "more wonderful than either steam packets or steam carriages." The engine would be able to move a huge pair of wings on either side of the horse.

Ada wrote her ideas to her mother, and said that she wished to also write them down in a book she wanted to call *Flyology*.

THE MECHANICAL AGE

In 1829, a Scottish historian wrote an important essay that called the early 19th century the Mechanical Age. Steam power was one of the major things invented during the Mechanical Age, which is why Ada wanted to use it in her flying machine. The Mechanical Age came before our Digital Age, which is filled with computer technology.

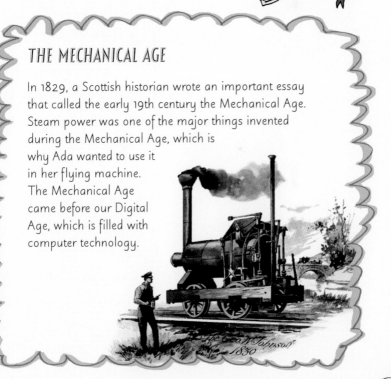

Annabella became impatient with Ada's flyology project. Ada was thinking about it a lot—too much in her mother's opinion. Annabella certainly didn't want to encourage Ada further by sending her a book about birds. Again, she feared that Ada was not spending enough time learning math.

Annabella wrote to her daughter to scold her for missing her studies. Besides, how could Annabella succeed in keeping Ada's feet firmly planted on the ground if she was always thinking about these fanciful things in the sky?

However, Ada did not feel the scolding was fair. She wrote back to her mother, explaining that she didn't think about flying when she was supposed to be concentrating on her lessons. Ada thanked her mother for her concern, but reassured her that she wasn't missing her studies. She then went on to say that she had decided to make smaller wings, which most likely angered Annabella even more, as Ada wouldn't give up on her project.

Ada's desire to fly could not be crushed. Although she refused to stop thinking about flyology, Ada couldn't control everything that was going on around her. Not long after Ada's 13th birthday, she became very ill. We can't be certain, but she might have had measles or polio—whatever it was, it made Ada very sick.

19TH-CENTURY ILLNESSES

Cities became overcrowded during the Industrial Revolution in the early 19th century. Many new factories were built that used new steam-powered machines. Lots of people moved to cities from the countryside to work. Living conditions became unclean, and diseases began to spread. Typhoid, polio, smallpox, measles, and cholera were just some of the diseases that could be caught in Britain during Ada's lifetime. At this time, it wasn't known what caused these diseases or how to cure them, and without access to modern medicine, many people died.

It took almost three years for Ada to recover from her illness. During that time she couldn't leave her bed and only had her schoolbooks to keep her entertained. She wrote in a letter that she had lost her "taste completely for riding and flying."

Anyone who has ever had to stay in bed for a long time knows that you have to find things to keep your mind busy. Ada dove into her studies.

PLANETARIUM

A planetarium is usually an indoor theater that recreates the night sky by projecting images of stars, planets, and constellations on the ceiling above you. As a science project, you can make a much smaller version of a planetarium out of a box—just like Ada did!

This artwork shows a 19th-century planetarium built by Italian astronomer Perini.

She could not move around but was well enough to spend the time reading and learning. She studied German, taught herself Latin, and made a planetarium out of a box.

Even though she was ill, Ada dedicated her time to becoming even more well-educated. But she also grew serious and no longer thought about flyology.

Did Ada lose her curiosity?

Presented at **court**

Once she regained her health, Ada began to blossom into an interesting young woman. She would soon enter English high society.

Now a teenager, Ada was smart, serious-minded, and curious about everything. Unlike her mother, she liked to wear brightly colored clothes.

Annabella decided that it was now time to broaden her daughter's education. She and Ada moved to Fordhook House, a large, white mansion next to the River Thames, which was very close to the hustle and bustle of London.

Annabella brought a portrait of Byron to Fordhook. She hung it

Fordhook House, where Ada lived with her mother, in Ealing, London. The house was once owned by the novelist Henry Fielding.

over the fireplace and covered it with a velvet cloth. The cloth was used to protect Byron's portrait from any light that might wear away the paint and ruin the painting. The portrait was shielded from everyone, but this didn't matter to Ada, as she claimed she didn't want to see it or even read her father's poetry.

Dr. William King

Dr. William King, one of Ada's tutors, frowned upon her reading Byron's poetry. He thought it would "lead to too much passion and a lack of self-control," traits that Byron was famed for possessing.

Interestingly, Ada's childhood dislike of math had vanished into thin air. Perhaps this change of heart was influenced by her mother and her numerous tutors. Or maybe Ada found math easy and she had finally embraced her talent for numbers.

Whatever the reason for this change, Ada found that numbers formed lovely shapes in her head as she solved problem after problem. She even preferred reading math books to any other books. Ada read complicated titles, such as Dionysius Lardner's *Analytical Treatise on Plane and Spherical Trigonometry*—probably not what her friends were reading!

Annabella enlisted the services of three of her friends, as well as tutors, to help manage Ada's education and also to prepare her for marriage. One of the friends was mathematician and scientist, Mary Somerville. Now 16 years old, Ada called her mother's friends the "Three Furies" because they hovered over her constantly.

MARY SOMERVILLE

Mary Somerville was born in 1780. Like most women growing up in the 18th century, Mary was not encouraged to get an academic education. Despite this, she read books and studied. Mary went on to become a mathematician, science writer, and astronomer. She correctly predicted the existence of the planet Neptune. Mary joined the Royal Astronomical Society in 1835 and was one of their first female members. She was a perfect role model for Ada.

King William IV   Queen Adelaide

When Ada was 17 years old, she was to be presented to the King and Queen at court. Being presented at court meant that Ada would meet the members of English high society in a formal ceremony. It was a major step in finding a suitable husband. Byron's famous daughter could now be courted, the 1830s term for dating.

There was a lot of preparation involved. Loads of time went into picking the right dress, the right accessories, and the right shoes for the ceremony. When the day finally arrived on May 10, 1833, Ada wore a white embroidered dress over a satin skirt. It had lots of ruffles and a train that flowed behind her as she walked. The dress was trimmed with feathers, pearls, and diamonds. Ada looked like a princess walking across the floor. Everyone was excited to see the daughter of Byron.

DEBUTANTES IN HIGH SOCIETY

The practice of presenting young ladies, called debutantes, at court started in the 18th century. This annual tradition signaled the start of the English social season, which was full of parties and balls. Ladies usually wore a white evening dress and white gloves. By 1958, the times had changed in Britain, and Queen Elizabeth II ended this practice. Today, debutante balls are still conducted in some parts of the United States.

Ada curtsied to King William IV and Queen Adelaide. Annabella was also in attendance, and she and Ada both met with dignitaries and famous guests. Ada was now accepted at court, considered an adult, and welcomed into high society.

Ada shared the news of her presentation to the King and Queen in a letter to her friend.

---

what is a dignitary?

A person who is considered to be important because they are high-ranking in government or in the church. Ada met dignitaries at court.

She wrote about how pleased she was with her mother during the event. Ada swelled with pride as she explained that Annabella looked very pretty, and she was even invited to sit next to the Queen!

Like other debutantes that were presented, Ada began meeting suitable men to marry. Tea parties, balls, horse races, and polo matches filled up her social calendar.

Life was becoming very exciting for Ada. She would soon meet people who would change her life forever.

Chapter 6

MEETING **Charles Babbage**

On June 5, 1833, Ada headed to Number 1, Dorset Street, near London's Manchester Square, for a party held by Charles Babbage.

Charles Babbage was a polymath. This means that he was an expert in lots of different subjects. Charles was a mathematician, philosopher, inventor, code breaker, and mechanical engineer.

Charles Babbage

Born in 1791, Charles was the son of a banker. Like Lord Byron, he studied at Cambridge University. After graduating in 1814, he gave lectures about a type of science called astronomy. Two years later, Charles was elected into the Royal Society of London, and soon became a professor of mathematics at Cambridge University.

Charles started working on a special invention to work out math calculations, called the Difference Engine, in the 1820s. When his father died in 1827, he inherited a large estate and great wealth. After his wife died in the same year, Charles paused his work and went to Italy.

On his return to England, Charles decided to open up his home to the great minds of the era, so he threw lots of parties. When Ada was invited to one in 1833, she was both excited and curious to attend a party at Charles's mansion near London's Manchester Square.

What is the Royal Society?

The Royal Society is the oldest national scientific organization in the world. Its members are recognized for excellence in science.

Party guests at Number 1, Dorset Street, often included author Charles Dickens, scientist Michael Faraday, mathematician Augustus De Morgan, scientist Charles Darwin, the Duke of Wellington, author Elizabeth Gaskell, astronomer Caroline Herschel, poet Alfred, Lord Tennyson, inventor John Herschel, and Mary Somerville. They were some of the brightest minds and most well-known people in the world. Many of them are still famous today for their accomplishments.

Ada's life had been somewhat quiet up until this point. Now she would be in a room with so many people that shared her interests—it was very thrilling.

Scientist Charles Darwin
made important discoveries
about evolution.

Author Elizabeth Gaskell
wrote stories about the rich
and poor people of England.

Astronomer Caroline Herschel
discovered many comets.

Scientist Michael Faraday
invented the electric motor.

Journalist Harriet Martineau reported that "all were eager to go to his glorious" parties. Everyone knew that you had to have at least one of three qualifications to be invited to Charles's parties: intellect, beauty, or status in society.

Harriet Martineau

When Ada arrived, one of the very first things that caught her eye was the silver dancer in the parlor—it was a mesmerizing clockwork toy. The dancer had a mechanical bird on her finger that could wag its tail, flap its wings, and open its beak.

LIKE CLOCKWORK

Charles Babbage wasn't the only person who had a clockwork toy. Centuries before Charles's silver dancer, Italian artist and inventor Leonardo da Vinci made a mechanical lion for King Louis XII, which could walk. Some of the most amazing clockwork toys, also known as automata, were created by Swiss watchmaker Pierre Jaquet-Droz in the 1700s. Three of his toys can still be seen in a museum in Switzerland.

But Ada's attention was also captured by something else in the parlor.

Charles Babbage had built a model of a machine that he called the Difference Engine. It was made out of thousands of carved cogwheels, and it worked a little like a calculator. This was unlike any other adding machine that had ever been produced, because after its switch was turned on it would work on its own—without any human help. It could work out huge math calculations.

There was only one problem—the machine was not finished. It was expensive to assemble, and there wasn't enough money to carry on building it. Construction had ground to a halt. However, groups of people, including Ada and her mother, were able to see the incomplete Difference Engine.

THE DIFFERENCE ENGINE

Charles's calculating machine, the Difference Engine, which had around 2,000 metal parts, can be seen at the Science Museum in London. A working model of the Difference Engine No. 2 was completed by the Science Museum in 1991. It was built to Charles's original design, and is made up of around 8,000 parts. It weighs 5 tons (4.5 tonnes) and is 10 ft (3 m) long.

Everyone watched Ada as she looked at the two-foot (just over half a meter) high, metal machine. Ada's friend Sophia, wife of mathematician Augustus De Morgan, noted that even though Ada was still so young, she understood how the Difference Engine worked. Sophia also wrote that Ada "saw the great beauty of the invention," while other guests looked at it with expressions of amazement, as if they were hearing a gun fired for the first time.

The other party guests could see Ada's interest in the machine, as she looked at it a lot, but no one could tell that her mind was spinning inside. This was like nothing she had ever seen before and it presented so many possibilities to her.

Charles was very impressed with Ada's intellect. It wasn't long before Ada and Charles began to write to each other about mathematical subjects. Ada had found a kindred spirit.

Like weaving flowers

In 1834, Annabella took Ada to the textile mills in northern England. There, Ada saw a machine that sparked her creativity.

Annabella wanted to find out about the lives of the textile workers. These workers had left their homes in the countryside, hoping to secure higher paying jobs in the new mills. But living conditions were bad both inside and outside of the factories.

THE INDUSTRIAL REVOLUTION

English working life began to change in the 1700s—there was a shift from farming to production on a large scale. Before this time, most manufacturing was done in people's homes using hand tools. People made their own clothes and produced their own food, and small village shops used only simple machines to make things. This changed when the steam engine was invented, as machines no longer had to be powered by hand. New factories were built in towns and cities, and lots of people moved from the countryside to get factory jobs. Across the ocean, the United States also began to experience a shift in industry, and mill towns grew there, too.

Whitchurch Silk Mill is the oldest silk mill in England, dating back to 1815. The mill can still be seen today in its original building in Hampshire.

The towns and cities that sprang up near the mills were not planned well. They couldn't support the number of people that flooded there to live. Living conditions had  become more and more unclean, and in 1831, a cholera epidemic killed hundreds of people.

Life inside the textile mills was just as bad. The workers faced long hours in dangerous conditions. New machines, which grabbed Ada's attention, had taken over the production of cloth and were extremely noisy. The air was filled with dust, and workers suffered from eye injuries, deafness, lung problems, and even cancer.

What is an epidemic?

When a disease affects a large amount of the population in a specific time period. The spread of cholera in 1831 was an epidemic.

During the Industrial Revolution, the textile mills in Britain were producing half of the world's cotton cloth. Not only was the work hard in these mills, but they also employed thousands of children, who faced the same bad conditions as the adult workers. There were many protests and strikes, but mill safety laws didn't come into effect until 1844.

It was no wonder that Annabella, who was interested in helping those less fortunate than herself, felt that she should visit the mills with Ada.

Ada saw the bad conditions in the mills, but she also noticed the new technology. She watched the Jacquard loom in motion. Like other looms, this loom used thread to weave patterns into cloth to create a design. However, the Jacquard loom was revolutionary. It was the first loom that used special cards with holes

THE JACQUARD LOOM

The Jacquard loom was invented in 1804. The system of punched cards it used was created in France by weaver and inventor Joseph Marie Jacquard. To demonstrate the system, a portrait of Jacquard sitting at his desk was woven in silk. This process used around 24,000 punched cards. The portrait was so detailed that it caught the attention of Charles Babbage.

A LA MÉMOIRE DE J. M. JACQUARD.

The silk portrait was woven in 1839 in memory of Jacquard, who died in 1834.

punched in them to form patterns in cloth. The Jacquard loom would inspire Charles to use similar punched cards to program math problems into his machines in the future.

Later, Ada wrote about the loom. She compared it to calculating machines, saying that they might weave "algebraical patterns just as the Jacquard loom weaves flowers and leaves."

This Jacquard loom is making cloth with a flower pattern.

HERMAN HOLLERITH AND IBM

Like Charles, Herman Hollerith, an American mathematician, was also inspired greatly by the Jacquard loom. In 1890, he used a similar system of punched cards to enter data into one of his machines. After Herman perfected his machine, he founded his own computing company, which he called the Tabulating Machine Company. Like any company, it had its ups and downs, but that changed when a man called Thomas Watson became the manager. Thomas thought it would be best to rename the company International Business Machines, which today is known around the world as IBM. IBM used Herman's method of punched cards for decades.

Herman Hollerith

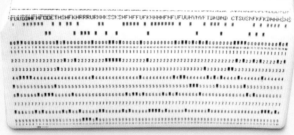

These punched cards have numbers "punched" out of them. IBM would have used similar punched cards to these.

Ada's imagination was sparked by the new technology she saw in the mills. She started thinking of technology more advanced than the Difference Engine, and began picturing what other machines might be able to do in the future. Ada thought that machines could be programmed to follow instructions, not just calculate the answer to math problems. In the future, she believed that machines could be used to create music and art—a truly groundbreaking idea. But first, she needed to understand how the Difference Engine worked.

Ada's visions were about machines that didn't exist yet—but her ideas are now seen throughout our Digital Age.

Chapter 8

The **Countess** of Lovelace

Charles's Difference Engine was very exciting. Ada was determined to study even more so she could fully understand how the machine worked.

Ada believed that the only way to get to grips with the Difference Engine was to concentrate solely on math. She knew that she needed to stop her thoughts from wandering elsewhere. Annabella encouraged Ada to pursue this goal by hiring Dr. William King to be Ada's tutor. Dr. King had one mission: to make sure that Ada was serious about her studies.

Ada liked her mother's choice of tutor and wrote directly to Dr. King. She told him that in order to focus her attention on scientific topics, she needed to stop thinking about other things. Ada had now realized that "nothing but very close and intense application to subjects of a scientific nature" could keep her imagination from running wild.

She dove into her studies with the same passion she had shown for her flyology project. While Ada was busy learning and attempting to stay focused, Charles Babbage was busy, too. He imagined how he could create another great machine—one that would be able to do much more than the Difference Engine.

As well as making sure Ada was studying hard, Annabella encouraged Ada to have a good social life. Annabella thought that Ada needed to be educated so that someday she would make an excellent wife. At the time it was believed that it was important for a woman to marry. Annabella also knew that for Ada to find a husband, she needed to meet more people.

When Ada was 19 years old, she was invited to a party where she met another man named William King. This William King was the 8th Baron King, and he was about 10 years older than Ada. He was considered very handsome and the two had similar interests—they clicked right away.

William King

William was a friend of a friend—Mary Somerville's son and William had both been educated by the same tutor. Mary knew that

William would be a great match for Ada. Annabella also encouraged this pairing. She liked that William had a title that had been passed down through generations, owned his own properties, and was a member of a family with high status in British society.

Ada and William were married in July of 1835, less than three months after they had met. Ada had her first child within a year of

TITLES IN ENGLISH SOCIETY

Some members of English society had titles in their names, such as Ada, Countess of Lovelace. There is an order, or rank, to these titles in English society. Many titles are hereditary—passed down from generation to generation. In 19th-century Britain, people with titles had power and wealth, and many titled people still do today.

King or queen

Prince or princess

Duke or duchess

Marquess or marchioness

Earl or countess

Viscount or viscountess, and baron or baroness

being married—a baby boy named Byron.

After Princess Victoria was crowned Queen in 1837, William climbed the ranks in English high society. He went from being a baron to a viscount, and then he earned the title of earl. William was then formally called the Earl of Lovelace. Ada also received a new title—Countess of Lovelace. She fully embraced this name change and even had calling cards (similar to modern-day business cards) made with her new name on. From now on, she would be known as Ada Lovelace.

This portrait of Queen Victoria was painted in 1843.

By 1839, the Lovelace family had two sons, Byron and Ralph, and a daughter, Anne Isabella, who was named after Ada's mother. Naming children after other family members was common at this time.

Ada's education didn't end when she got married. In fact, William encouraged his wife to spend her free time studying.

Ada wrote a letter to Mary Somerville saying that she was reading math books every day and was currently in the middle of learning about algebra and trigonometry. Ada made it clear to Mary that even though she was married

what is trigonometry? The study of triangles. Ada was learning about trigonometry in her math book.

"So you see that matrimony has **by no means** lessened my taste for these pursuits, nor my **determination** to carry them on."

Ada Lovelace
in a letter to
Mary Somerville

with three young children, she still wanted to continue her math lessons with another tutor.

Augustus De Morgan, a respected mathematician, became Ada's new tutor. He taught Ada her now favorite subject—math. However, he wrote to Ada's mother and husband expressing his concerns about teaching Ada such a tricky subject.

Augustus
De Morgan

Augustus was worried about women's ability to study advanced math. He thought that the complicated subject would put too much strain on their bodies, resulting in bad health. Augustus believed that only men were strong enough to study math because the subject needed to be tackled with a strength beyond "a woman's physical power." His way of thinking might seem strange to us today, but it fitted in with the views of the time. However, Ada proved this theory wrong. She was a talented mathematician, and she wrote an essay in which she explained that curiosity and imagination are needed to make new discoveries in math and science.

Augustus saw Ada's creativity and mathematical ability. He encouraged her to study hard, and Ada never gave up on her goal of understanding the Difference Engine. Some people might have thought that Ada's questions were unladylike, but Ada didn't care—she continued to work hard and her learning grew deeper.

The Analytical Engine

Ada and Charles wrote to each other often. They shared a common love of mathematics, inventions, and science.

Charles had now set aside his Difference Engine for an even more remarkable invention—the Analytical Engine. This steam-powered machine would be formed of two parts: the "mill" and the "store." These parts would work in a similar way to the processor and memory in modern computers. The mill would do math calculations, and the store would keep a record of any calculations that were done. Charles thought numbers could be entered into the machine by using punched cards.

What are punched cards?

These cards have a pattern of holes, or "punches," which represent numbers that can be "read" by a computing machine.

WORLD'S FIRST COMPUTER

Often called the "Father of Computers," Charles Babbage designed the world's first computer—the Analytical Engine. He first wrote about this machine in 1837. It was the successor to his Difference Engine, and it introduced computing concepts that are still used today. Although it was never fully built, Charles worked on it until he died in 1871.

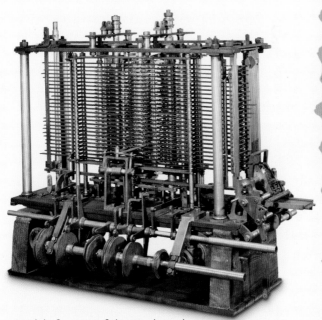

Model of a part of the Analytical Engine in the Science Museum in London, England.

Charles spent a good deal of time making drawings and notes, but he did not have a way of actually creating his Analytical Engine. The parts he needed hadn't been invented.

While building the Difference Engine, Charles had worked with an English toolmaker called Joseph Clement. Joseph had made some of the parts for the previous machine, but the size and complexity of this new design were too much for him to take on. The Analytical Engine would have to be powered by steam in the same way as the Difference Engine, but it was to be a lot bigger than the first machine. Charles decided that the design needed more work.

Ada, Annabella, and Mary Somerville were invited to Charles's home. Charles shared the plans for his new machine, and the four talked well into the evening. Charles told them that he was close to figuring out how to build his new

machine, but he didn't have the materials or tools. Charles said that he felt like he was at the top of a mountain, waiting for the mist to clear in the valley below to reveal something—his Analytical Engine.

Ada was amazed by his idea, and could see the machine's potential even while it was just plans on paper. When finished, the machine would be enormous, about the size of a steam train, and it would be able to do lots of things, such as work out math problems. But Ada knew this was going to be so much more than just a calculating machine.

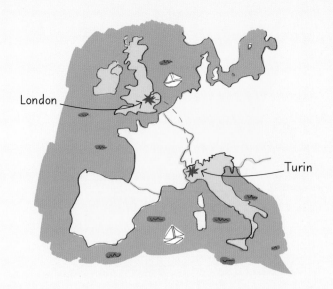

London

Turin

In 1840, Charles was invited to travel from London to Turin, Italy, to explain his Analytical Engine in a lecture.

Before he left, Charles had written to his Italian host, Count di Plana, about his high hopes for the lecture. He was confident that his machine was too advanced to be made in England and would be better understood and appreciated in Italy.

One of the people who attended the Turin lecture was an Italian mathematician and military engineer called Luigi Menabrea. Luigi was an extremely intelligent young man, who later became the prime minister of Italy. He would have heard Charles's talk and saw the drawings of the Analytical Engine.

Impressed, Luigi published his account of the invention so that others could read about it. His 1842 paper was called *Notions sur la Machine Analytique de Charles Babbage* ("Ideas on the Analytic Engine of Charles Babbage"). It was printed in French in a journal published by the *Bibliothèque Universelle de Genève* (Universal Library of Geneva).

At the same time, Ada continued her studies. Before Charles's trip, she had even begun teaching math to her own students—two women who were daughters of her mother's friends.

Portrait of Luigi Menabrea in 1885

"I get so **eager** when I write **mathematics** to you that I forget all about handwriting and everything else ..."

Ada Lovelace in a letter to her students

In a letter to her students, Ada explained that she was really looking forward to teaching them math. Concentrating on her students' progress in lessons was very exciting to Ada. So much so that she'd often forget how to do other things, even how to write neatly!

As well as sharing her wisdom with her math students, Ada would soon share her thoughts, ideas, and visions with the rest of the world. She decided she would translate Luigi Menabrea's text about Charles's machine into English.

Chapter 10

The enchantress

Ada agreed to translate Luigi's text from French to English. She decided to add her own notes on the Analytical Engine to the translation.

Once the translation was complete it would be published in the journal *Scientific Memoirs*.

As Ada worked on the text, she noticed a few things that needed to be explained a bit more. Charles thought she should write her own paper about the machine, but Ada was reluctant to. This wasn't a common thing for a woman to do in the 1800s, especially for someone of Ada's social status. So instead, Charles suggested she include her own notes along with Luigi's text.

Ada wrote and wrote. In "Note A," Ada's longest note, she introduced the possibility that numbers input into the Analytical Engine could output something other than numbers. Ada thought the machine might be able to produce colors, for example. She then wondered what else it might be able to do if it was ever completed.

One of Ada's notes, "Note G," was more important than all the others. In it, Ada wrote that the Analytical Engine might be able to work out a special series of numbers—known as the Bernoulli numbers—by itself.

BERNOULLI NUMBERS

In the 18th century, Swiss mathematician Jacob Bernoulli first studied what became known as Bernoulli numbers. These are a special set of numbers that play an important role in math. The numbers are "recursive." This means that the first number is used to calculate the second, the first and second to calculate the third, and so on. They are still used today.

Jacob Bernoulli

She said that the machine could work out these numbers, instead of humans having to do the hard work. The method of working out the Bernoulli numbers explained in "Note G" is called an algorithm. This algorithm became known as the first computer program.

During the translating and note writing, Ada and Charles wrote many letters to each other. The mail delivery system in the 19th century was very different from how it is today.

Ada and Charles sent handwritten letters to each other around six times a day. These were carried by a servant between Ada and Charles's homes, which were about a mile (1.6 km) apart. It was almost as easy for them as it is to send a text message today. Ada would write a question to Charles on notepaper, have it delivered, and he would answer quickly with another letter.

While working on the translation, Ada had other responsibilities. She was a mother, a wife, and a member of high society. Her days were filled with activities, such as playing with her children and attending society events. In one letter she told Charles that she couldn't get

any work done because she had too much to do. Ada was a very busy person. But, like other women of her social status, Ada had servants that handled many household tasks.

This meant that she still had lots of time to study. Even while Ada was taking care of her family and enjoying other activities, such as horseback riding, she was still thinking about the Analytical Engine. Like anyone who has a project deadline looming, it was always on Ada's mind. She thought about it constantly.

It took Ada around six months to complete the translation along with her notes. This was even with her husband, William, helping her make copies of the text. When she was finished, her notes far outnumbered Luigi's original text by roughly 20,000 words.

In 1843, Ada's translation of Luigi's text and her notes were published. Its title in the *Scientific Memoirs* journal was "Sketch of the Analytical Engine Invented by Charles Babbage" by L. F. Menabrea. Ada's name didn't appear in the journal, but her initials,

A. A. L., followed each note she wrote. Ada knew that this was a great achievement.

Despite spending most of her life trying not to be creative, Ada looked at the Analytical Engine in a poetic, creative way. Ada believed Charles's machine could be so much more than just a time-saving machine.

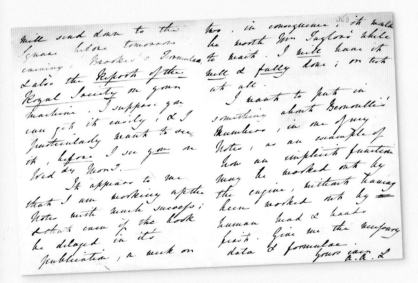

This letter from Ada to Charles describes her idea for an algorithm that could be used to work out math problems using the Analytical Engine instead of by hand.

What is an algorithm?

A step-by-step method that is used to solve math problems. Ada had ideas for algorithms that the Analytical Engine could use.

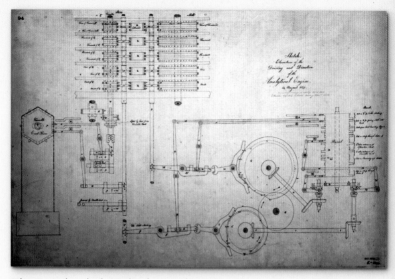

This is a sketch that Charles made in 1841. It shows some of his plans for the Analytical Engine.

She even thought that "the engine might compose elaborate and scientific pieces of music" in the future.

Unfortunately, Charles's machine was never built, so Ada couldn't test it out. But this didn't stop her from thinking about its potential. Charles was impressed with Ada's ideas. He nicknamed her the "Enchantress of Numbers," as he felt that her math abilities were almost magical. He wrote that she understood scientific concepts in a way that most men in Britain couldn't.

"[Ada has] thrown her **magical spell** around the most **abstract of Sciences** and has **grasped** it with a force which **few masculine intellects** (in our country at least) could have exerted over it."

Charles Babbage, writing about Ada

English artist Margaret Carpenter painted this portrait of Ada in 1836.

Ada's priority was not just to be known as the author of her notes. Instead, she wanted to write well to make sure she would have the opportunity to write more about the Analytical Engine in the future, using her initials again.

When the translation was finally printed, Ada was so happy with it that she sent finished copies to her mother. William was extremely proud of his intelligent wife. He gave copies of the work to his friends, claiming that it would help to boost his family's standing in English society.

Chapter 11

The Great Exhibition

The success of Ada's translation and notes made her want to become a science writer like her friend Mary Somerville.

Ada thought of her notes as her written children. She wanted to continue writing, and had time as she had nannies for her three actual children.

Ada had other things on her mind, too. She sent a letter to Charles in which she suggested that they work together on building the

Analytical Engine. Ada proposed that she take on a role similar to a modern-day CEO and Charles should be like a modern-day CTO.

Ada even joked in her letter that Charles would have to take orders from a woman. Ada was so excited by the prospect of working with Charles that she told her mother she was going to be in charge of the project, calling herself the "High Priestess of Babbage's Engine."

However, Charles decided not to work with Ada. He felt that they had different opinions about how the project should be done. Also, the machine was going to cost too much money to build. At the same time, Ada's health began to fail. She had a disease called cancer and couldn't spend a lot of time working.

Charles approached the British government, asking for money. They had already given him money to build

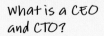

What is a CEO and CTO?

CEO stands for Chief Executive Officer, who is the person in charge of a company. CTO stands for Chief Technology Officer, who is the person in charge of a company's technology.

the Difference Engine, which was never finished, so they refused his request.

Charles decided that he could raise funds by charging people money to see a mechanical tic-tac-toe game he had designed. However, the machine was never built.

Then Charles came up with the idea that perhaps Queen Victoria's husband, Prince Albert, could help. Prince Albert was very interested in the arts, trade, industry, and science. He wanted Britain to be the world leader in innovation in technology. Prince Albert had always been curious about scientific

HOW TO PLAY TIC-TAC-TOE

Tic-tac-toe is a game played by two people. One person draws Os, and the other draws Xs, in a grid. The aim of the game is to draw three Os or Xs in a row before the other person does. Charles wanted to build a mechanical version of this game.

inventions, and Charles's were no exception. Charles wanted to use this interest to persuade the prince to give him money to build the Analytical Engine—but he never actually asked the prince. To make matters worse, Charles was not invited to Prince Albert's Great Exhibition in London in

Prince Albert

1851. Charles had fallen out with the British government over money and was not allowed to show off his incomplete Difference Engine, even though it was a brilliant example of mechanical engineering.

The Great Exhibition was also known as the Crystal Palace Exhibition, named after the glass and iron building that it was held in. Prince Albert came up with this idea so that people from around the world could show off their inventions. Even Ada's husband, William, took part, winning an award for brick making.

The Great Exhibition was a huge fair that featured new technology from across the globe. Between May and October of 1851, it was visited by more than 6 million people.

SILK & VELVE

COTTON

HOROLOGICAL
INSTRUMENTS

PHILOSOPHICAL
INSTRUMENTS

LONDON
TIME

After the Great Exhibition, Ada's health worsened. By 1852, she was less active with each passing day. Ada had been touched by a death scene she had read in a book by well-known author Charles Dickens. Her husband, William, asked Charles Dickens to read to her on her deathbed, to prepare her for death.

In November 1852, Ada lost her fight with cancer. She passed away, leaving behind her mother, her husband, and her three

Charles Dickens

children. Even though she barely knew her father, Lord Byron, Ada was buried next to him at the Church of Saint Mary Magdalene in the English town of Hucknall in the county of Nottinghamshire.

Florence Nightingale, the founder of modern nursing, wrote about Ada after she died, praising her genius. She said that Ada "could not possibly have lived so long, were it not for the tremendous vitality of the brain, that would not die."

Inscribed
by the express direction of
Augusta Lovelace
[...] 10th 1716: Died [...] 27th 1352
to recall her Memory.

If faith shall save the sick, [...] shall raise him up
committed sins, [...] shall be forgiven him
James 5 : 1 5

[...] in hope, in thankful [...] or who mourn
that boundless work of [...] Soul [...]

[...]TED BY HER MOTHER A.F NOEL BYRON MDCCCLIII

This memorial of Ada can
be found near where she
grew up in Leicestershire.

Chapter 12

A LEGACY OF **poetical science**

Ada's notes on the Analytical Engine, and her ideas about what it could do in the future, are why she is known as the first computer scientist.

Before Ada passed away, she wrote a long letter to her mother. She told Annabella that she wanted to use the creative genius her father, Lord Byron, had "transmitted" to her to make a meaningful contribution to the world of math, science, and technology. Ada wanted to share her ideas and knowledge with everyone.

Ada had practiced what she called "poetical science" throughout her work, as she combined her intelligence and determination with her curiosity and imagination.

what is a legacy? Something that someone is known for doing that impacts the future. Being the world's first computer scientist is Ada's legacy.

This gave her the ability to think about what might be possible in science in the future, instead of just seeing what was in front of her. A lot of Ada's ideas about technology were ahead of their time, such as thinking that machines might be able to compose music or that they could discover math concepts that weren't obvious to humans.

While Ada's vision of what future machines could do was exciting, she also saw their limits. When discussing the Analytical Engine with Charles, Ada argued that machines would never be able to think for themselves. She was certain that computers would not be able to create

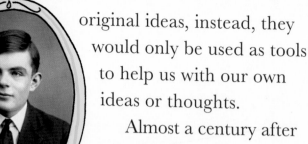

original ideas, instead, they would only be used as tools to help us with our own ideas or thoughts.

Almost a century after Ada died, a math and science genius called Alan Turing developed the Turing Test.

Alan Turing

Alan created this test to find out if computers could think for themselves. In the Turing Test, a human asks a "partner" a set of questions. Using the answers, the human has to decide if the "partner" is a computer or not. In 2014, Ada's theory, which Turing had named "Lady Lovelace's objection," was proved wrong when a computer that was programmed to answer the questions like a 13-year-old boy passed the test.

ALAN TURING

Alan Turing was born in Britain in 1912, 60 years after Ada died. He was a very clever man who knew lots about math and science. During World War II (1939–1945), Alan was a code breaker who worked out messages that were sent by Britain's enemy in the war, Germany. He had ideas for how to create programmable machines, and made a design for a computer. It was called the Automatic Computing Engine (ACE), but it was never built—like the Analytical Engine. Alan's contributions to science were groundbreaking, similar to how Ada's were a century earlier.

Alan's design of the ACE computer is seen here in the Science Museum in London.

Over the years, Ada's legacy started to fade. However, in the late-1970s, the United States' Department of Defense (DOD) created a computer language named after her. This language is used to program computers around the world, which are used in different industries, such as in the military, airlines, railways, and banking.

However, Ada's legacy isn't just about her visionary work with early computers. She is also a role model for all girls interested in learning about science and math. Ada grew up in a time when women were expected to be good wives and mothers and not focus on their education—this made her achievements in science so momentous. She defied what was expected of her, was determined to learn as much as she could, and constantly proved wrong the people who doubted her abilities.

Fortunately, times have changed, and in today's world, STEM (science, technology, engineering, and math) careers are widely available to both men and women. Most high-ranking jobs in these industries are still done by men, although women are strongly encouraged to enter into these fields.

STEM

STEM is a term that stands for science, technology, engineering, and math—all the things that Ada enjoyed studying. Nowadays, these four subjects are closely linked, and STEM topics are taught together by using real-world examples. In the 19th century when Ada was alive, it was widely thought that STEM subjects were only of interest to men, but of course this is not true. Many women enjoy learning about science, technology, engineering, and math.

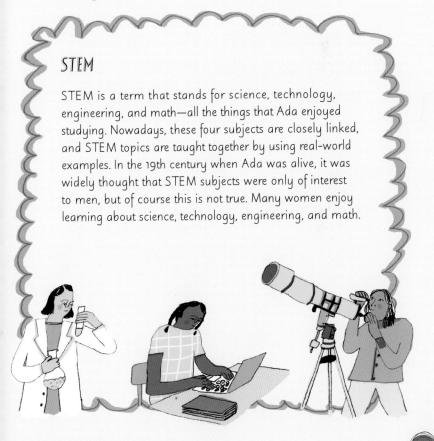

Because of Ada's influence, there are now many women who have been able to make important discoveries, build incredible inventions, and make a difference to the world of science.

Women today are recognized for their academic achievements. Every year, on the second Tuesday of October, people around the world celebrate women who work in STEM on Ada Lovelace Day.

Ada would love that, don't you think?

Ada's
family tree

Grandfather

Captain John "Mad Jack" Byron
1756–1791

Grandmother

Catherine Gordon
c. 1764–1811

Aunt

Augusta Maria Leigh
1783–1851

Father

Lord George Gordon Byron
1788–1824

Augusta was Byron's half sister. They had the same father, but different mothers.

| Grandfather | Grandmother |
| Ralph Milbanke 1747–1825 | Judith Noel 1751–1822 |

Mother

Anne Isabella "Annabella" Milbanke
1792–1860

Ada Lovelace
1815–1852

Husband

Lord William King
1805–1893

↑ Ada married William in 1835.

Son
Byron King-Noel
1836–1862

Daughter
Anne Isabella King
1837–1917

Son
Ralph Gordon King-Milbanke
1839–1906

Timeline

On December 10, Ada is born.

In April, Ada writes to her mother to say that she has worked out how to make wings.

1815 **1816** **1824** **1828** **1829**

Around a month after Ada is born, her parents separate.

Ada gets sick. She will stay in bed recovering for the next three years, keeping herself busy with her studies.

Byron dies fighting in the Greek War of Independence against Turkey.

Ada tours textile mills in northern England with her mother and sees the Jacquard loom.

In May, Ada is presented at court to King William IV and Queen Adelaide of Britain.

1833

1834

1835

Ada marries William King.

In June, Ada meets Charles Babbage and sees his incomplete Difference Engine for the first time.

Ada gives birth to her first child, Byron.

Charles is invited to give a lecture on his Analytical Engine in Turin, Italy.

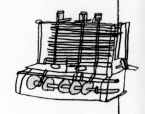

Ada gives birth to her third child, Ralph.

Ada gives birth to her second child, Anne Isabella.

1836 1837 1838 1839 1840

Charles first writes about his Analytical Engine, which introduces some computing concepts that are still used today.

William becomes an earl and Ada becomes a countess.

The United States Department of Defense creates a computer language in Ada's honor.

On November 27, Ada dies.

Ada's translation of Luigi's text is published in English.

1842   **1843**   **1851**   **1852**   **1970s**

Italian mathematician and engineer Luigi Menabrea publishes a paper written in French about Charles's Analytical Engine.

Prince Albert holds his Great Exhibition, in London. Charles is not asked to show his Difference Engine, but Ada's husband, William, wins an award for brick making.

Quiz

1   Why was Ada's father famous?

2   What nickname did Byron give Annabella?

3   What was the name of Ada's Persian kitten?

4   How did Ada imagine her flying horse would be powered?

5   How old was Ada when she was presented to the king and queen at court?

6   What was the first thing that caught Ada's eye at Charles Babbage's house?

7   What inspired Charles to use cards with holes punched into them to program math problems into his machines?

**Do you remember what you've read?
How many of these questions about
Ada's life can you answer?**

8   Why did Ada have to have tutors instead of studying at a university?

9   What is the name of the machine Charles worked on after the Difference Engine?

10   In the notes Ada added to her translation of Luigi Menabrea's text about the Analytical Engine, how many more words did she write than him?

11   Between May and October of 1851, about how many people visited the Great Exhibition in London?

12   When is Ada Lovelace Day celebrated?

Answers on page 128

Who's who?

Adelaide
(1792–1849) queen of
Britain from 1830 to 1837

Albert
(1819–1861) Queen
Victoria's husband

Anning, Mary
(1799–1847) English
fossil collector

Babbage, Charles
(1791–1871) mathematician,
philosopher, inventor,
code breaker, and
mechanical engineer

Bernoulli, Jacob
(1655–1705) Swiss
mathematician who studied
what became known as
Bernoulli numbers

Byron, George Gordon
(1788–1824) Ada's father,
the well-known poet

Darwin, Charles
(1809–1882) English
scientist who made
important discoveries
about evolution

De Morgan, Augustus
(1806–1871) mathematician
who was one of Ada's tutors

De Morgan, Sophia
(1809–1892) Ada's friend,
married to Augustus
De Morgan

Dickens, Charles
(1812–1870) English author
of famous novels, such as
A Christmas Carol

Faraday, Michael
(1791–1867) English
scientist who helped to
invent the electric motor

Frend, William
(1757–1841) mathematician
who was one of Ada's tutors

Gaskell, Elizabeth
(1810–1865) English author
who wrote stories about
rich and poor people

Herschel, Caroline
(1750–1848) German
astronomer who discovered
many comets

Hollerith, Herman
(1860–1929) American mathematician and inventor who was inspired by the Jacquard loom

Jacquard, Joseph Marie
(1752–1834) French inventor of the Jacquard loom

King, Anne Isabella
(1837–1917) Ada's daughter

King, Lord William
(1805–1893) Ada's husband

King, William
(1786–1865) one of Ada's tutors

King-Milbanke, Ralph Gordon
(1839–1906) Ada's younger son

King-Noel, Byron
(1836–1862) Ada's older son

Leigh, Augusta
(1783–1851) Byron's half sister and Ada's aunt

Martineau, Harriet
(1802–1876) English journalist who once reported on Charles Babbage's parties

Menabrea, Luigi
(1809–1896) Italian mathematician and military engineer who later became prime minister of Italy

Milbanke, Anne Isabella "Annabella"
(1792–1860) Ada's mother

Mitchell, Maria
(1818–1889) American astronomer who discovered a comet

Nightingale, Florence
(1820–1910) English nurse who worked to make hospitals cleaner and safer

Somerville, Mary
(1780–1872) Scottish mathematician, astronomer, and science writer

Turing, Alan
(1912–1954) English mathematician and computer scientist who created the Turing Test

Victoria
(1819–1901) queen of Britain from 1837 to 1901

William IV
(1765–1837) king of Britain from 1830 to 1837

Glossary

academic
relating to school or university

algorithm
step-by-step method used to solve math problems

anatomy
study of the body of a human or animal

automata
clockwork toys

Bernoulli numbers
special set of "recursive" numbers, where the first number is used to calculate the second, the first and second to calculate the third, and so on

cancer
disease caused by cells that are not normal, which can grow and spread throughout the body

CEO
chief executive officer, the person in charge of a company

code breaker
someone who works out secret coded messages

computer scientist
person who studies computers

Continent, the
all the countries in mainland Europe in the 19th century; this didn't include the islands in Europe

courted
dated, with the goal of marriage

CTO
chief technology officer, the person in charge of a company's technology

debt
money that someone owes to another person or a bank

debutante
young woman who is presented to society

dignitary
person who is considered to be important because they are high-ranking in government or in the church

epidemic
disease that affects a large number of people in a specific time period

heiress
woman who will inherit wealth, property, and status from her family or other person

hereditary
passed down from generation to generation

House of Lords
one part of the British government

Jacquard loom
first weaving loom that used cards with holes punched in them to form patterns in cloth

legacy
something that someone is known for doing that impacts the future

lift
upward force that allows birds to fly

Luddites
group of English workers in the 19th century who protested against new technology in the textile mills where they worked, which was costing them their jobs

marvel
something that inspires amazement

mesmerizing
fascinating

parallelogram
four-sided shape with opposite sides that are the same distance apart along their whole length

planetarium
indoor theater that recreates the night sky by projecting images of stars, planets, and constellations on the ceiling

polymath
expert in lots of different subjects

punched cards
cards with a pattern of holes, or "punches," which represent numbers that can be read by a computing machine

reputation
opinion that people have about a person, or what a person is known for

Romanticism
style of art and literature that values emotion and imagination

Royal Society, the
oldest national scientific organization in the world

social season
time of year, usually during winter, where members of high society can meet people they might marry

society
exclusive group of wealthy
and fashionable people

status
someone's rank or
position in society
compared with others

steam packet
type of boat that is
powered by a steam engine

STEM
science, technology,
engineering, and math

successor
person or thing that
comes after someone
or something else

trigonometry
study of triangles

visionary
person able to imagine
likely possibilities for
the future

vitality
lively energy of spirit

Index

Aa

Ada Lovelace Day 108
Adelaide, Queen 46, 48–49
air pressure 35
Albert, Prince Consort 96–97
algebra 72
algorithms 86, 89
Analytical Engine 76–85,
    88–97, 102–103, 105
anatomy 36
Anning, Mary 17
astronomy 17, 40, 45, 50
automata 54
Automatic Computing
    Engine (ACE) 105

Bb

Babbage, Charles 50–57,
    62–63, 67, 76–84, 86–91,
    94–97, 103
Bernoulli, Jacob 85
Bernoulli numbers 85–86
birds 12, 34–38
Byron, Captain John 19
Byron, Lord George Gordon
    8–12, 19–25, 27–28,
    42–44, 50, 100, 102
Byron, Lord William 19

Cc

calculating machines 55–57,
    63, 79
calling cards 71
Cambridge University 17, 19,
    20, 50–51
cancer 60, 95, 100
Carpenter, Margaret 92
celebrities 8, 12, 18, 20, 24, 33
CEO (Chief Executive
    Officer) 95
Childe Harold's Pilgrimage 12
childhood 8–16, 26–31
cholera 39, 60
cities 39, 60
Clement, Joseph 78
clockwork toys 54
clothes 42, 46–47
clouds 12
colors 13–15, 85
comets 17, 53
computers 37, 76–79, 86,
    103–107
constellations 40
Continent 32–33
court, royal 46–49
Crystal Palace Exhibition,
    London 97–100
CTO (Chief Technology
    Officer) 95

Dd

Darwin, Charles 52, 53
De Morgan, Augustus 52, 74–75
De Morgan, Sophia 56, 75
debutantes 47–49
Department of Defense (US) 106
di Plana, Count 80
Dickens, Charles 52, 100
Difference Engine 51, 55–57, 65–67, 75–78, 96, 97
Digital Age 37, 65
dignitaries 48
diseases 39–40, 60, 95, 100
DOD computer language 106

Ee

education 13–19, 27–31, 44–45, 72–75
electric motors 53
Elizabeth II, Queen 47
English Channel 32
epidemics 60
Europe 32–33
evolution 53

Ff

factories 39, 58–65
Faraday, Michael 52, 53
Fielding, Henry 43
flying 34–39, 41, 67
Fordhook House, Ealing, London 42–43
Frend, William 13–15

Gg

Gaskell, Elizabeth 52, 53
Great Exhibition, London 97–100
Greek War of Independence 24

Hh

heiress 18
Herschel, Caroline 52, 53
Herschel, John 52
Hollerith, Herman 64
horse, flying 36–39
House of Lords 20, 21

Ii

illnesses 39–40, 60, 95, 100
Industrial Revolution 39, 59–61
Italy 24, 51, 80–81

Jj

Jacquard, Joseph Marie 62
Jacquard looms 62–64
Jaquet-Droz, Pierre 54

Kk

King, Anne Isabella 72
King, Dr. William 44, 66–67
King-Milbanke, Ralph 72
King-Noel, Byron 71
Kirkby Hall, Leicestershire 26, 30

Ll

Lamb, Lady Caroline 21
Lamont, Miss 27, 30
languages 19, 27, 28–29, 41, 106
legacy 102
Leigh, Augusta 23
Leonardo da Vinci 54
letter writing 28, 86–87
looms 62–64
Lovelace, William King, Earl of 68–72, 88, 93, 97, 100
Luddites 20, 21

Mm

mail delivery system 86–87
marriage 18–19, 21–23, 49, 68–74
Martineau, Harriet 54
measles 39
Mechanical Age 37

Menabrea, Luigi 80–81, 83–84, 88
Milbanke, Lady Anne Isabella "Annabella" 8–12, 18, 21–24, 26–39, 45, 58, 61, 68–69, 102
Mitchell, Maria 17
music 27, 28, 33, 65, 90, 103

Nn

nannies 26–27, 94
Nightingale, Florence 17, 100
numbers 44–45, 85–86

Oo

Oxford University 17

Pp

parallelograms 22–23
Perini 40
planetariums 40–41
planets 40, 45
"poetical science" 102–103
poetry 10, 12, 20, 31, 43–44
polio 39
polymaths 50
Puff, Mistress 26–27
punched cards 62–64, 76
punishments 28, 31

Rr

rainbows 12–15
recursive numbers 85
reputation 10
Romanticism 12
Royal Astronomical Society 45
Royal Society of London 50, 51

Ss

Saint Mary Magdalene, Hucknall, Nottinghamshire 100
Science Museum, London 56, 77, 105
Scientific Memoirs 84, 88–89
sky 12–15
smallpox 39
social season 21, 47
Somerville, Mary 45, 52, 68–69, 72–74, 78
steam power 36–37, 39, 59, 76, 78
STEM subjects 107–108
sunlight 14–15
Switzerland 32, 54, 85

Tt

Tabulating Machine Company 64

technology 20, 21, 65, 95–99, 103, 107
Tennyson, Alfred, Lord 52
textile mills 21, 58–65
"Three Furies" 45
tic-tac-toe 96
titles 70
trigonometry 72
Turin 80–81
Turing, Alan 104–105
Turing Test 104
tutors 13–15, 17, 19, 26–30, 44–45, 66–67, 74–75
typhoid 39

Uu

United States 59, 63
universities 17, 20, 67

Vv

Victoria, Queen 71, 96

Ww

Watson, Thomas 64
weaving 62–64
Wellington, Duke of 52
Whitchurch Silk Mill, Hampshire 59
William IV, King 46, 48
wings 34–35, 38–39
World War II 105

Acknowledgments

DK would like to thank: Rebekah Wallin for proofreading; Hilary Bird for the index; Maya Frank-Levine for the reference section; Allison Singer and Shannon Beatty for editorial help; and Stephanie Laird for literacy consulting.

The author would like to thank: the DK team, including Katie Lawrence and Marie Greenwood, the staff at the Morgan Library, and her agent, Jennifer Laughran. She also wants to thank all of the amazing teachers who engage young women in STEM exploration.

ANSWERS TO THE QUIZ ON PAGES 116–117

1. he was a poet; 2. Princess of Parallelograms; 3. Mistress Puff; 4. by steam engine; 5. 17 years old; 6. the silver clockwork dancer in the parlor; 7. the Jacquard loom; 8. as a woman, she wasn't allowed to go to university; 9. the Analytical Engine; 10. 20,000; 11. more than 6 million; 12. the second Tuesday in October